MCR

EVANSTON PUBLIC LIBRARY
W9-BJM-582

Talking About Time

Months of the Year

Angela McHaney Brown

Raintree
Chicago, Illinois

Published by Raintree, a division of Reed Elsevier, Inc.
Chicago, Illinois

Customer Service 888-363-4266
Visit our website at www.raintreelibrary.com

For information, address the publisher:
Raintree, 100 N. LaSalle, Suite 1200, Chicago, IL 60602

Printed and bound in China by South China Printing Company Ltd.

09 08 07 06 05
10 9 8 7 6 5 4 3 2 1

Library of Congress Cataloging-in-Publication Data
A copy of the cataloging-in-publication data for this title is on file with the Library of Congress.
Months of the Year / Angela McHaney Brown
ISBN 1-4109-1642-1 (HC), 1-4109-1648-0 (Pbk.)

Acknowledgments
The publisher would like to thank the following for permission to reproduce copyright material: Alamy Images p. 17; Corbis pp. 12, 15; Corbis pp. 6 (L. Clarke), 9 (Sandy Felsenthal), 21 (Gary Hurewitz), 11 (Ronnie Kaufman), 19 (Jose Luis Pelaez, Inc.), 13, 18 (Ariel Skelley); Getty Images p. 16; Getty Images p. 22 (Imagebank); Harcourt Education Ltd pp. 4, 5, 7, 8, 10, 14, 20, 23 (Tudor Photography).

Cover photograph reproduced with permission of Harcourt Education Ltd (Tudor Photography).

Every effort has been made to contact copyright holders of any material reproduced in this book. Any omissions will be rectified in subsequent printings if notice is given to the publisher.

Some words are shown in bold, **like this**. You can find out what they mean by looking in the glossary on page 24.

Contents

Months of the Year

A **month** has about 30 days.

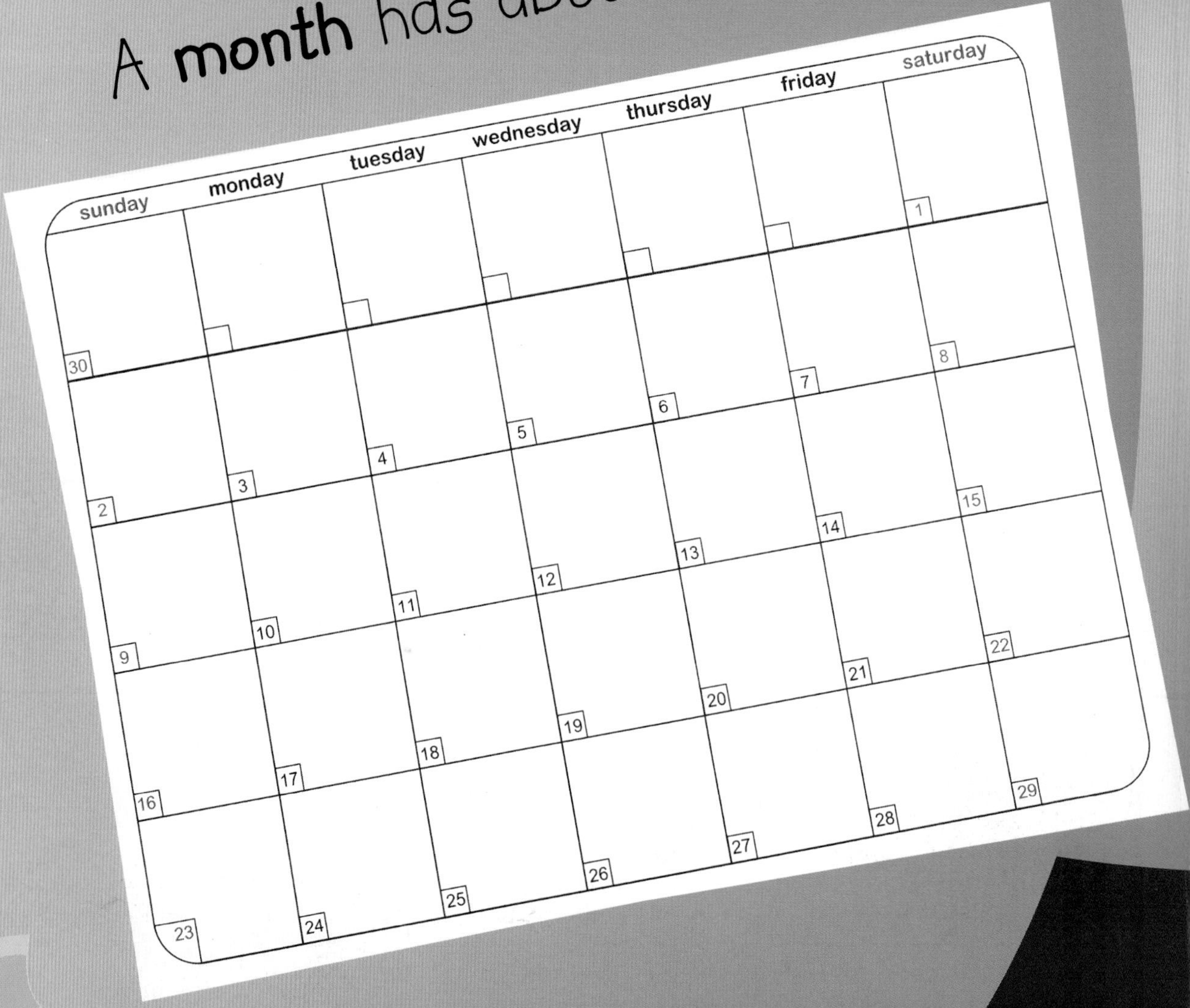

There are 12 months in a **year**.

Do you know what they are?

1 January
2 February
3 March
4 April
5 May
6 June
7 July
8 August
9 September
10 October
11 November
12 December

A New Year

New Year's Eve is the end of the **year**.

January

January 1 is the first
day of a new year.

February

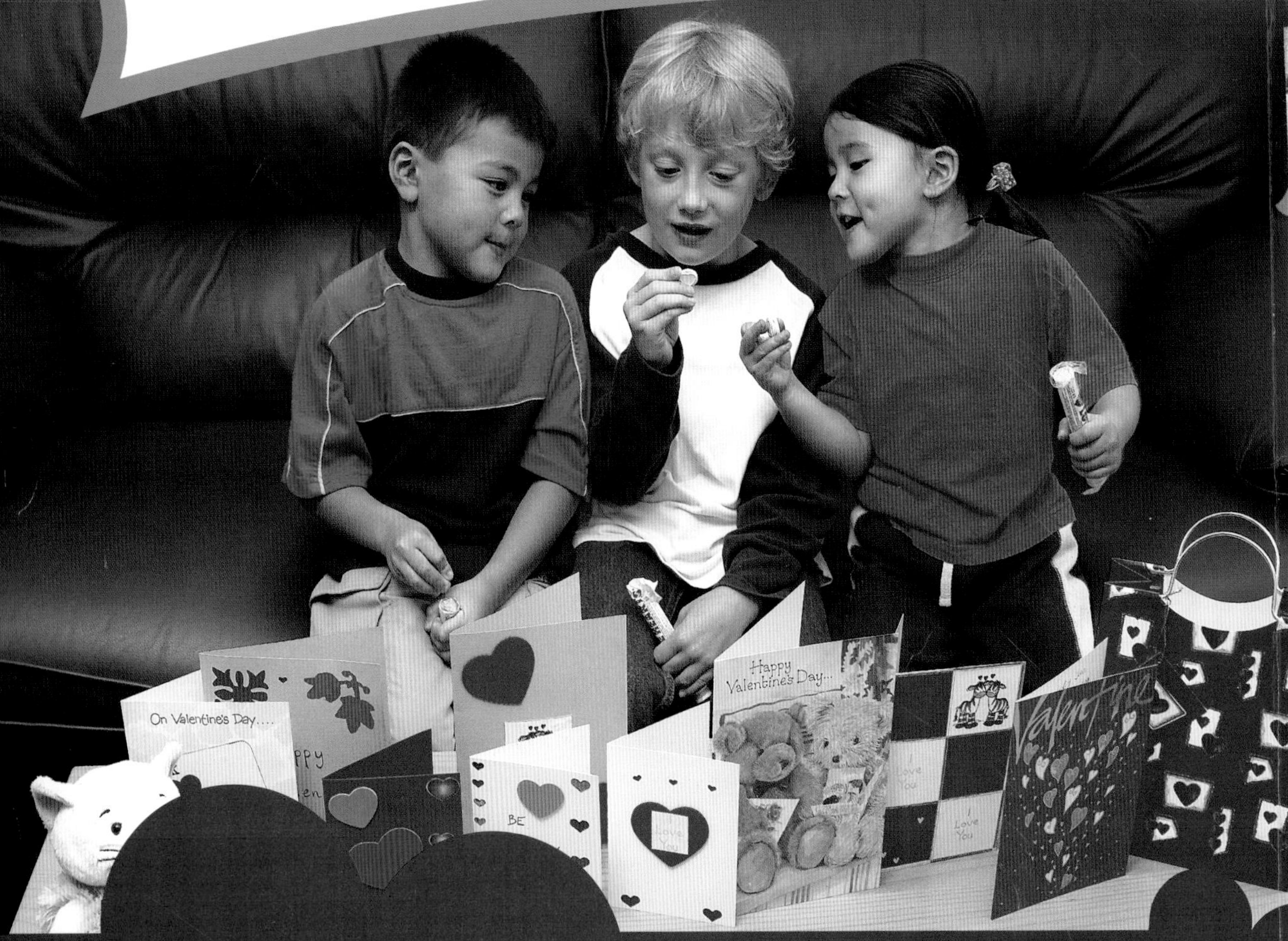

Julie and her friends eat
candy hearts on Valentine's Day.

March

Maggie wears green on St. Patrick's Day.

April

Bobbie plays a funny joke on her brother.

May

Jane plants flowers with her mother.

Summer is almost here!

June

The last day of school is here!

Sophie can't wait to play with her friends all **summer**.

July

The **year** is half over now.

July 4 is Independence Day.
Hurray, fireworks!

August

Sara goes swimming. What do you do in the **summer**?

September

Jen goes back to school.
She will make new friends.

October

Some leaves turn different colors. Does this happen where you live?

November

Katie's family
has a turkey
on Thanksgiving.

December

December is the last **month** of the **year**.

The **holidays** are here.

Do you celebrate a holiday?

Happy New Year!

December 31 is **New Year's Eve.**

Tomorrow starts another new **year**!

Glossary

April Fools' Day April 1, when people often play jokes on each other

month amount of time equal to about four weeks or 30 days; there are 12 months in a year

New Year's Eve December 31, the last day of the year

summer often warmer season between spring and fall, usually including June, July and August

year about 365 days, which is the time it takes the Earth to go once around the Sun

Notes for adults

The Talking About Time series introduces young children to the concept of time. By relating their own experiences to specific moments in time, the children can start to explore the pattern of regular events that occur in a day, week, or year.

This book focuses on the twelve months of the year and shows what children might see and do in each of the twelve months. The use of recognized holidays helps to reinforce how some activities happen at a predictable time each year. As the months are presented chronologically, the book will encourage children to observe the cyclical nature of our years and build up a sense of how long a year is.

Follow-Up Activity
Write down each of the twelve months on a separate note card. Mix up the cards and ask the child to place them in the correct order, while discussing what might happen during each month.

Index